A UNIVERSE OF STORIES

Deborah S. Wallace

ISBN: 978-1-7342049-3-3
Library of Congress Control Number: 2020904937

All photos courtesy NASA (www.nasa.gov) except where noted.

Cover design by Judith S. Design & Creativity
www.judithsdesign.com
Published by Glass Spider Publishing
www.glassspiderpublishing.com

I dedicate this book to those teachers who over the years have inspired me, and for their selfless dedication and hours devoted to me and my classmates, helping us to love learning and believe in ourselves. The teachers that come to mind are my third-grade teacher, Mrs. Chase; my fifth-grade teacher, Mrs. Greenberg; my high school band teacher, Mr. Faulks; my flute teacher, Victoria Scoones; my college social policy professor, Emma Gross; and my school social work professor, Margie Egbert. My favorite teacher is Donna Wallace, my sister, who dedicates endless hours of her personal time to provide a quality education for all her students.

But most of all, I want to dedicate this book to my parents, Elmore and Ruth Wallace, who have been my lifelong teachers and mentors. They have always encouraged me to pursue my education and have led by example. My dad, a retired IBM engineer and my mother, an IBM executive secretary who became a stay at home mom, have never lost their love for learning and adventure. My dad, who will strike up a conversation with anyone, has learned much about people and our connection with each other as human beings. My mom is an adventurer, and at the age of 82 is up for anything—a huge storage building fire in her town, and she's running down the street with her camera in hand to get closeup pictures of the firefighters in action, or she is playing the pipe organ for the town's Holiday celebration, or taking care of a sick neighbor half her age. Both have taught me it's not just the learning, but what you do with that knowledge to make the world a better place.

Deborah S. Wallace

TABLE OF CONTENTS

INTRODUCTION

The goal of this book is to inspire youth and the young at heart to explore more about the STEM information mentioned in these stories—to spark one's curiosity to learn more about the scientific ingenuity and technology that advances our knowledge about our universe, our world, and ourselves.

May you be inspired to write your own stories about the universe and other science topics, and to bring together art and science as you expand your knowledge and creativity together with lifelong learning and curiosity.

Always foster your innate curiosity and maintain your childlike spirit of adventure.

A small glimpse of our universe that shows about 5,500 galaxies in the Hubble Ultra Deep Field.

THE BIG BANG THEORY OF COSMIC EVOLUTION

Imagine traveling back to a formless and empty beginning,
Our cosmos not yet visible to us.
Being much smaller than a pinpoint size
Yet containing all manner of matter and energy
That may ever and forever henceforth exist.

It was a dark period with no time, no space, no stars shining.

Looking through a special microscope,
You could possibly view this subatomic world.
Like looking at a drop of water
Through a microscope in biology class,
Watching amoebas slithering about.

All quantum forces joined in brief unity—
Extreme dense subatomic energy
Spawning heat greater than the Sun's core.
Under extreme pressure, it bursts forth.

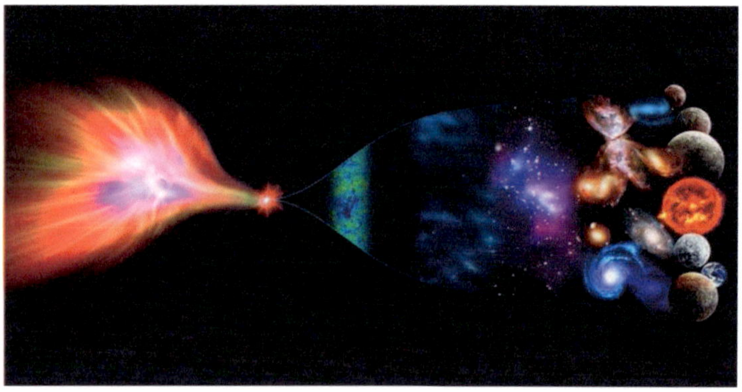

With light invisible to the naked eye, our big bang began.
Much faster than light speed, our universe gave form.
Just as our night sky backdrop appears dark,
Special telescopes give way to rich colors.

Forces in motion, our destiny beckons,
Divinely creating space and time as one.
Space-time was nonexistent until that very instant,
Expanding in all directions—rapid inflation.

Fractions of a second
Doubling hundreds of times.
Tiny fireball spewing forth
From nothingness to a small marble in size.

Charged subatomic particles wildly swirling.
Matter battling anti-matter particles.
Matter always winning by just one.

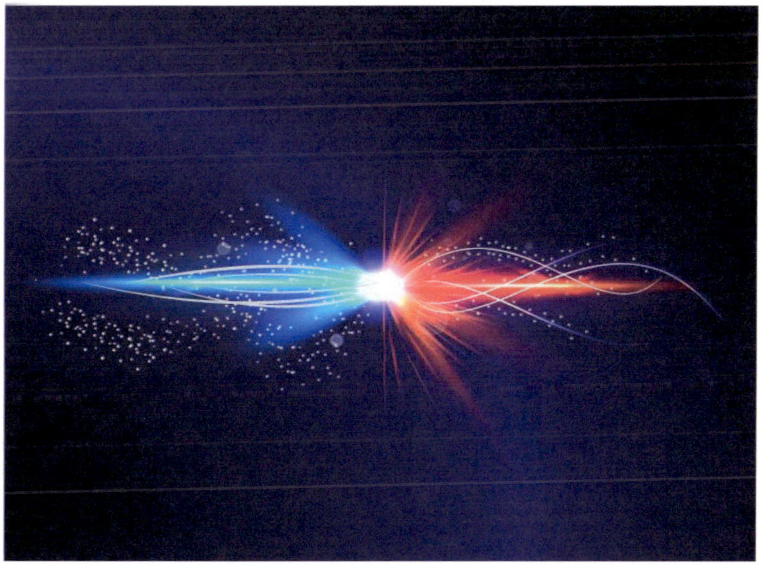

Energetic photons bounce off free electrons,
Stopping light photons from traveling freely,
Like sunrays scatter as storm clouds gather,
Lost within the storm's dark fury.

Thick, opaque particle soup cools as it expands.
Protons and neutrons combine, the first atoms now forming.
Three hundred eighty thousand years since the big bang—
A brilliant flash, transparent light shines briefly through.

Still no stars in these dark ages; dark space still abounds.

This brilliant flash was dubbed the "cosmic microwave background."
This once infrared light and its echo still remain—
Cooled microwaves, electromagnetic afterglow
Filling all cosmic space; a background relic, leftover remnant.

Cosmologists' predicted chance discovery,
A mysterious hum in a radio antenna.
Microwaves' frigidness, absolute zero,
Impossible to see with the naked eye.

Some of television's between-channels noise and snow
Show static and white noise of the microwave background.

Microwave telescopes unveil vintage portraits,
Revealing the grand cosmic background.
A skyscape imprint, an almost unvarying image,
Like a slightly smudged fingerprint surviving for eons.

Inflating, cooling, then slowing,
Dense spots of matter and gas gather.
Dense matter seed clusters—reds, yellows, and blues—
Like paint colors splashed on a canvas background.
Gas cluster seeds clump; matter together grows.

Our future stars, planets, and galaxies.

Matter bending space-time with a force called gravity.
Space-time inflation causing the expanse of clusters.
Stars, planets, and galaxies forming faster.
Gravity giving galaxies their orbits.

Our universe began with rapid expansion.
Cosmic inflation uniformly expanding space
Like the surface of a hot air balloon taking form,
Continuing ever outward, 11.4 to 13.8 billion years strong.

Our universe for a time cooled and slowed,
Gravity binding galaxies together.
Galaxies now speed forward, faster and faster,
With a mysterious force called dark energy.
All galaxies moving farther and farther away from us,
Space stretching, carrying all matter, galaxies, and people with it.

One day, the cosmic light of galaxies will not reach us; dark skies will enshroud us.
With recorded history our only reference point, we'll be alone in dark space
Like old pictures of long-deceased relatives we will never meet.

Inflation is like hooks on two tow trucks
If the universe had an edge.
Imagine each hook pulling opposite sides of the universe,
Stretching more and more as the tow trucks drive away.

Gravity—like cables on a winch,
If you could secure cables around the entire universe
As the winch winds up the cable,
Everything in the universe would be pulled inward.

Inflation versus gravity:
Space-time's ultimate tug of war.
The winner determines our final fate.
Calculations predict a big freeze, big rip, or big crunch end.

A big freeze if the hooks and chains barely win—
Slightly stronger than the winch cables,
Cables stretched to sagginess.

The universe and galaxies pull apart,
All suns and planets float away,
Alone in dark, cold, frozen space.

A big rip end if the hooks and chains are mightiest—
Snapping the winch cables like twigs,
Galaxies and solar systems tear apart.
Everything, even atoms and subatomic particles, destroyed.

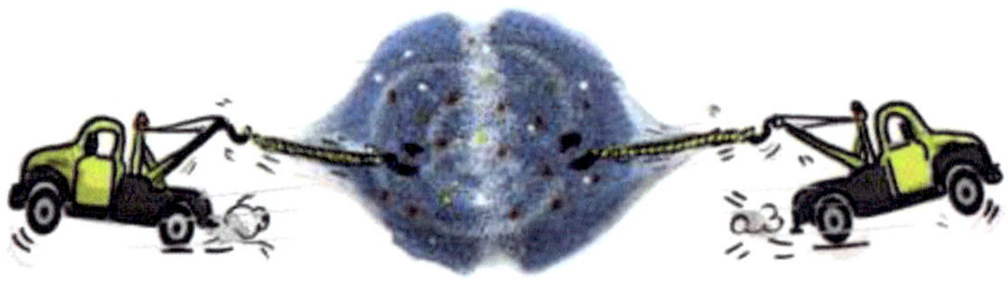

A big crunch end if the winch cables win—
As gravity brings the expanding universe to a stop,
Pulling the universe together.

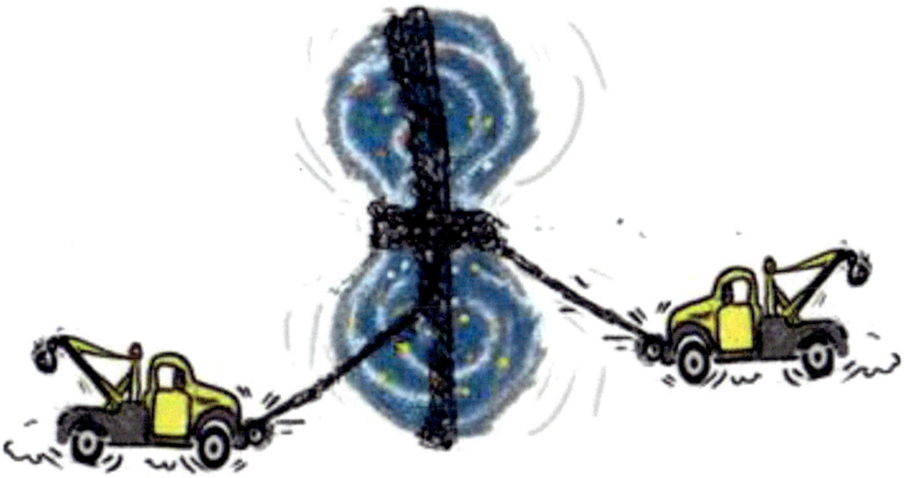

Galaxies rush toward each other, colliding.
Temperatures rise, everything compresses.
Atoms and subatomic particles crunch together,
Creating an incredibly hot, dense, compact universe.

Or will the end be the beginning of another universe?
Not general relativity's singularity theory
But a quantum subatomic cyclical big bounce.

Leftover remnants of a dying universe,
Space-time crunched to a subatomic speck,
Once again exploding,
Reigniting a big bounce.

Expanding, contracting, collapsing, repeating.
Rising from the dust of its predecessor
As the great Phoenix rising from the ashes.

Infinite space-time's life cycle:
Birth, death, rebirth.
As space-time controls the cosmos' evolution from its birth,
So space-time will control how it dies and evolve.

Our universe was possibly not created alone—
Bubble universes may exist all around us
With laws of physics differing from our own
In a multiverse of many other universes.

Exactly how a subatomic quantum world
Evolved into the general relativity universe of today,
And what existed before the big bang,
Is still an unsolved mystery—leaving the cosmic door open
As other theories and the universe evolve in tandem.

The graphic below shows a timeline of the universe based on the big bang theory and inflation models.

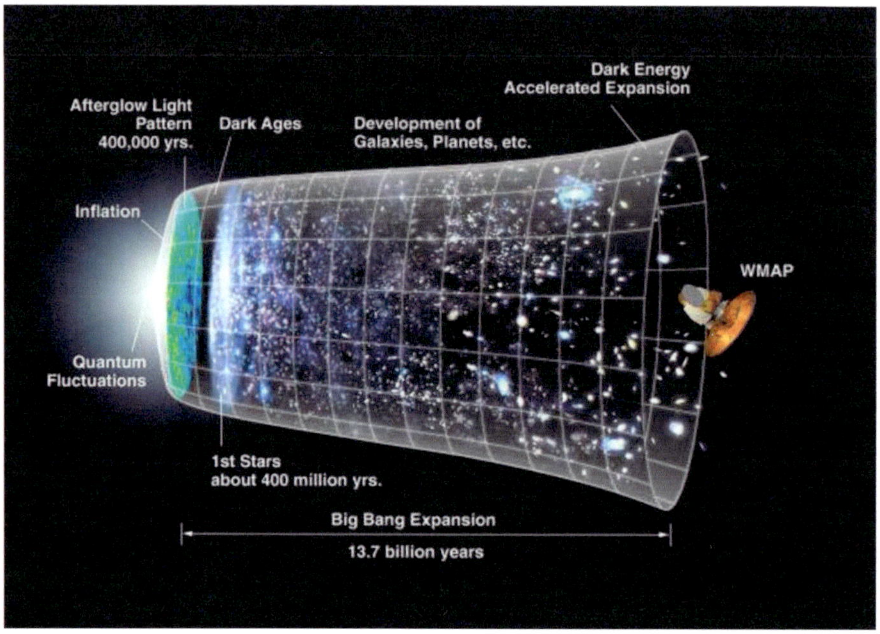

The image below shows changes in the rate of expansion since the big bang.

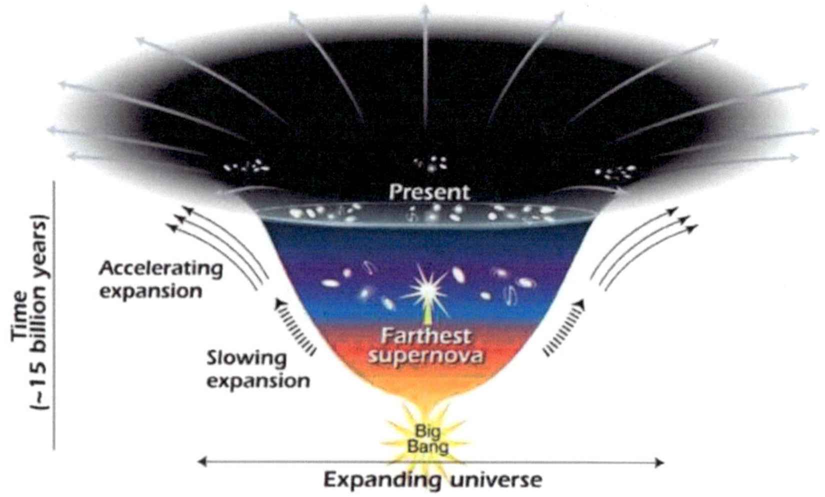

The cosmic microwave background (CMB).

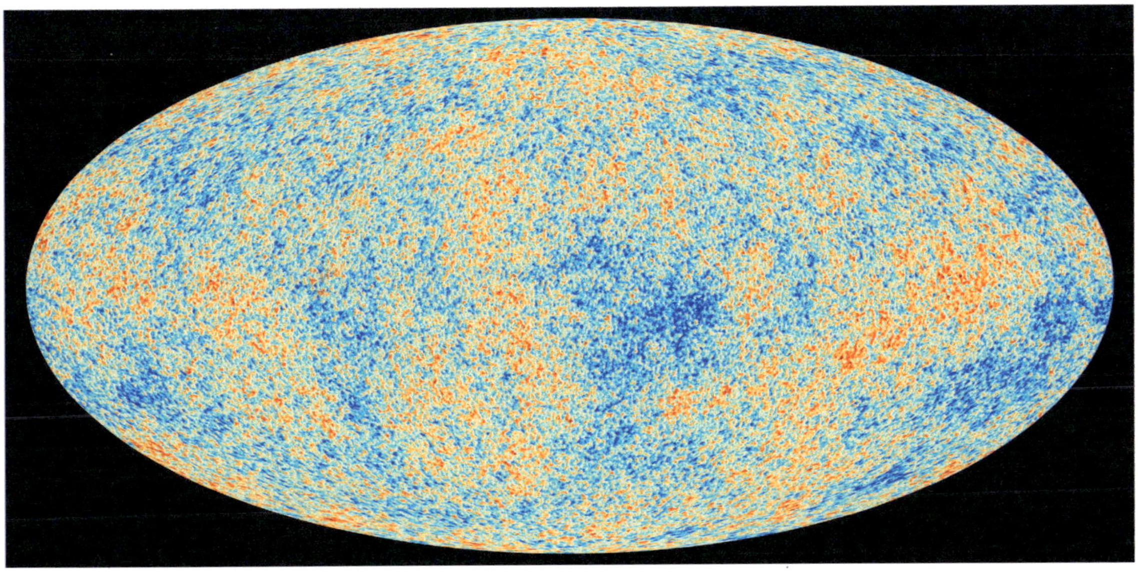

DARK ENERGY

Dark Energy: the name sounds so ominous.
A pressure lurking beneath the galaxies.
A mysterious force acting opposite gravity,
Accelerating against gravity's pull.

Traveling back in time
One hundred years ago,
The Milky Way galaxy imagined as the entire universe,
Existing as a universe of stars
One hundred thousand light-years across.
Standing still, neither growing nor shrinking.

Albert Einstein writes his math formulas on a chalkboard,
Equations to explain his universal theory of gravity.
His formulas calculate an expanding or contracting universe.

Yet believing the universe unchanging and eternal,
He adds the Greek symbol Lambda, an upside-down capital V—
Einstein's cosmological constant.

Its sole math function, opposing gravity's pull.
Pushing outward against gravity's natural desire
To draw the entire cosmos into one giant mass.

Like holding your forearms up straight
Shoulder width apart, hands facing each other
As your friends try to push your arms together.
Your resistance is what Lambda represents:
Stopping gravity's universal collapse.

Human "computer" Henrietta Leavitt,
Studying stars called Cepheid variables,
Imaged on glass photographic plates as tiny black dots.
In a time before camera film or computers.

We too are "computers" when using only a pencil and paper
To add, subtract, multiply and divide—
Computing calculator-free.

Henrietta discovered these stars pulsate reliably over time,
Dim to bright, then repeating.
Cepheid stars, the cosmos' first "standard candles."

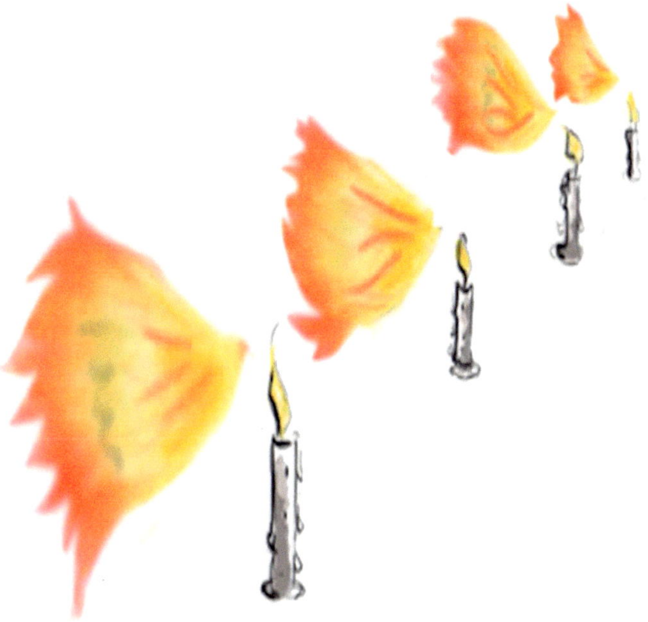

A historic first—now we're able to measure stars' real distances.
Like guessing we're almost to the mountain peak—
It's just around the next bend—
When we actually have two more miles to hike.

High on a California mountaintop
Sits a newly created, far-reaching telescope.
A one hundred inch mirror
Magnifying distant starlight.

Edwin Hubble peers through this telescope
At far-distant spiral nebulae,
And clouds of interstellar gas and dust,
Seeing Cepheid variable stars.

Finding the cosmos extends
Far beyond the Milky Way's reach.
Fuzzy nebulae once thought to be
patches of gas and dust,
Now unique "island" galaxies,
Each with 100 billion stars.

Andromeda galaxy, our closest neighbor,
Over 2 million light-years away.

Edwin Hubble also measured stars' speeds,
Revealing they travel outward
As the universe expands.
Lambda's cosmological constant
Is now Einstein's biggest error.

Starlight radiates wavelengths,
Displays a rainbow of colors through a prism.
Shorter blue wavelengths move toward us;
Longer red wavelengths move away.

Like a passing siren's wail,
Sound wavelengths compress and stretch.
The siren sounds loud and fast coming toward you,
Wavelengths compressed:
AWhoo, AWHoo, AWHOo, AWHOO.

The siren fades, sound slows moving away.
Wavelengths stretched:
A...WHOoo, A....Whoooo, A......whoooo, a......whoooooooo.

When passed through a prism,
Most starlight in the universe shines red
As galaxies move away from us.
More distant ones move away faster
As the universe and space itself uniformly expand.

Galaxies remain intact, held together by gravity
As space carries galaxies along with it.
Like kayaking downstream in a swift and fast river.
The river takes you and your kayak along with it
No matter how hard you try to paddle upstream.

Like evenly placed raisins in bread dough—
Baking, the dough rises and moves the raisins with it.
Sitting on your raisin galaxy while your friends sit on theirs,
Each sees the other move farther and farther away,
Creating the illusion that your galaxy isn't moving at all.

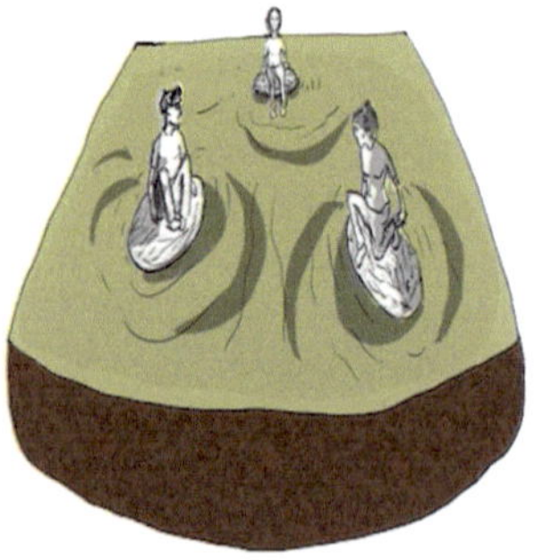

One hundred years later,
Lambda, Einstein's cosmological constant symbol,
Pushing outward against gravity's force,
Is added back into universal equations,
Now given a new value in calculating this accelerated expansion.

Lambda, representing a repulsive energy force,
Once believed to be Einstein's biggest mistake,
Today is called dark energy
And is believed to be this constant,
Always present throughout the cosmos' history.
Dark energy may best explain the expanding universe.

Measuring supernovae explosions, a new standard candle
Deepens our understanding of dark energy's power.
Exploding stars show the universe speeding up.

Dark energy's presence is seen only by its effects.
Like a windy day, tree branches bending to and fro
By unseen wind rustling fallen leaves,
Blowing them down the road.

Invisible dark energy pushes cosmic expansion.
Dark energy's mysterious antigravity pressure,
Negative vacuum energy of empty space,
Pushing rather than pulling,
Existing throughout time.

Like using the blower side of the shop vacuum
Instead of sucking up debris,
It now blows air out, pushing the debris away.

Not measured in any lab,
Its exact properties are an enigma.
Empty space is not obligated to show its true makeup.

Dark energy may expand exponentially into the future.
New technological advances,
Like the 100-inch telescope in Hubble's day,
Will help solve this dark matter mystery.
High on a Chilean Andes mountaintop,
A five hundred seventy megapixel digital camera

Mounted on a far-reaching telescope
Maps out hundreds of millions of galaxies.
Detecting thousands of supernovae explosions.
A worldwide scientific collaboration called the Dark Energy Survey.

It measures dark energy changes and any slowing over time,
Just as the early universe expanded faster than the speed of light
And then slowed down as it cooled.
Dark energy is predicted to slow down once more.

The Dark Energy Survey and future scientific advances
Continue to fill in more pieces of this dark energy puzzle.
Dark energy hides like the mysterious figure in a Magic Eye picture.

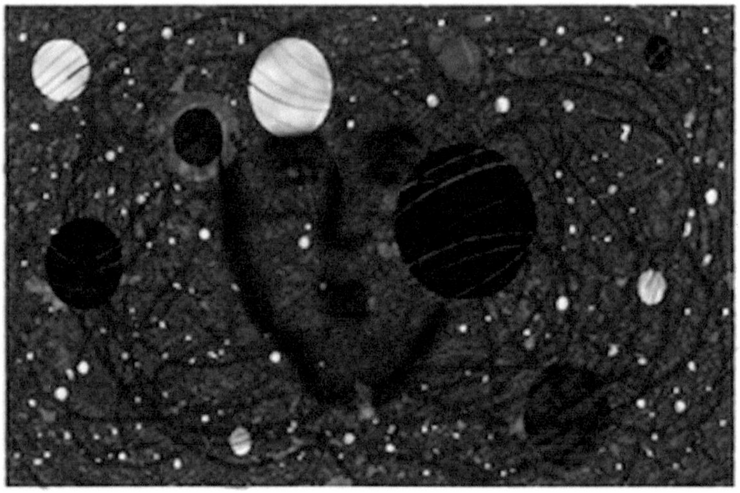

Dark energy ruling 75% of the universe's energy:
21% from dark matter, 4% from normal matter.

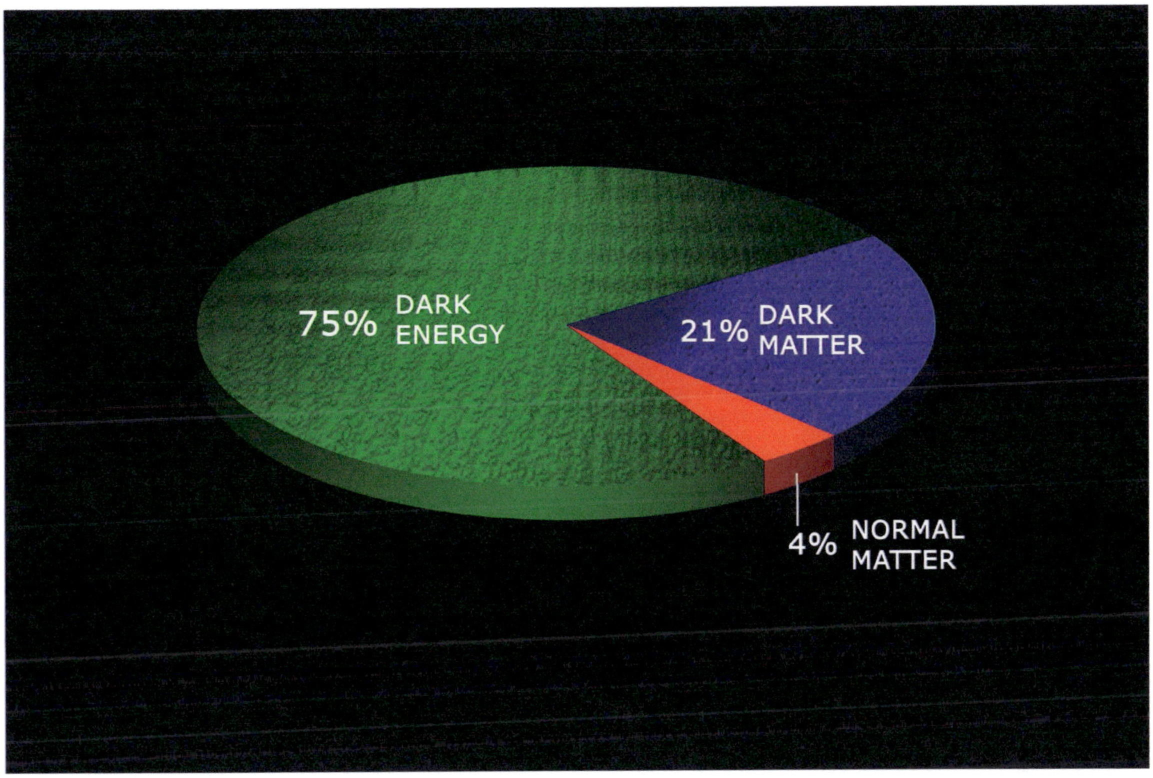

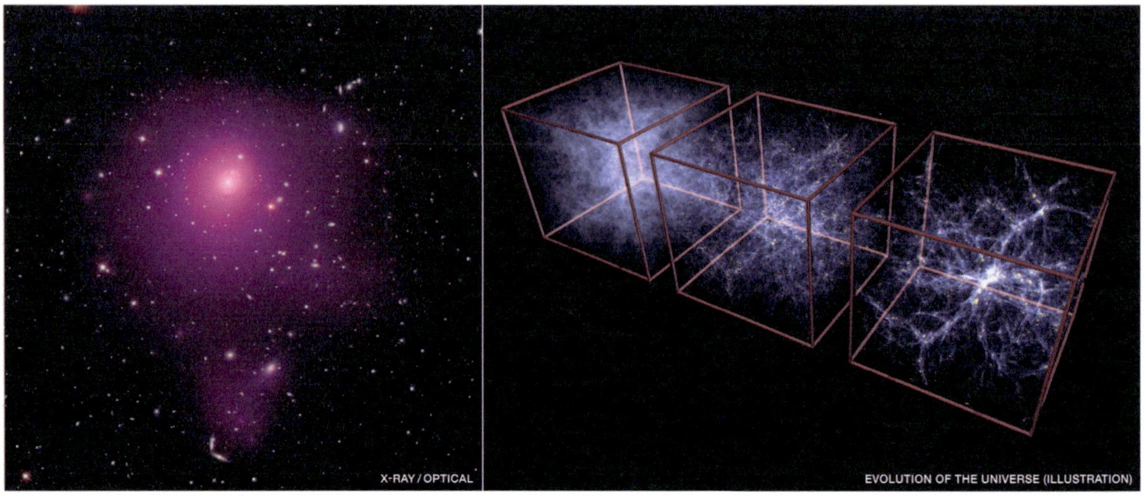

The image on the left is of the galaxy cluster Abell 85. The purple gas detected in X-rays by NASA's Chandra X-ray Observatory and the other colors show an optical image of galaxies from the Sloan Digital Sky Survey. This galaxy cluster observed by Chandra shows how dark energy has limited the growth of the galaxy cluster over billions of years.

The image on the right shows snapshots from a simulation that represents the growth of the structure of the universe 0.9 billion, 3.2 billion, and 13.7 billion years old. Growth was initially produced by the gravity, but later interrupted from competition from dark energy's repulsive force.

GRAVITY

Gravity: inescapable and universal.
Grounding us to Earth,
Stopping us from floating away.

Gravity always works on everything with mass:
Particles, people, planets, galaxies.

A roller coaster climbs up and over the track's top,
For an instant suspended in air,
A feeling of brief weightlessness.

Cars free fall under gravity's force,
Velocity increasing, coaster and people flying down the tracks,
Speeding faster and faster, stopping safely
Slowed by air currents, the track's friction, and good brakes!

Galileo imagined dropping objects of varying mass
From atop of the Leaning Tower of Pisa.
In the absence of air currents,
Metal balls and feathers land simultaneously.

Discovery: under gravity's influence,
All objects free fall, all with the exact same acceleration,
Each second of time falling faster and farther.

Place a piece of paper on your notebook.
When dropped, the notebook and paper touch
down together,
Together falling faster no matter the height—
Galileo's law of free fall.

Isaac Newton sits under an apple tree
Watching apples fall to the ground.
The Moon in the sky above
Wonders how far-reaching Earth's gravity may be,
Attracting falling apples and the Moon in orbit around it.

Newton's math formulas calculate this attraction,
Showing any two objects always pull on each other in the same way.
Gravity's force long-range mutual attraction knows no limits.
Gravity's calculated strength lessens over distance.

Large celestial objects' tugging power is greater than smaller ones.
The Moon's weaker pull affects our ocean tides.
Earth's stronger tidal pull locks the Moon, slowing its rotation.
The Moon travels a straight line,
Falling slightly sideways toward Earth.
Earth's gravity pulls the Moon's travel into orbit around it.

Cannonballs shot outward
Curve before hitting the ground.
Newton calculated cannonballs hurled fast and hard enough
Would launch into orbit around Earth.

Gravity's pull decreasing over distance.
Closer Mercury speeds around the Sun (112,000 mph),
Distant Neptune moves much slower (12,146 mph).

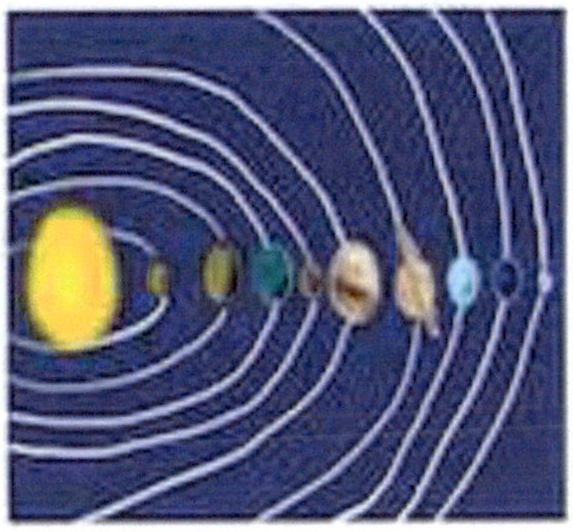

Gravity's invisible force
Holds our solar system together,
Known only by its attractive effects.

Yet gravity is the weakest force of nature.
Take a small magnet—paperclips cling to it.
Magnetic power out-muscles planet Earth's
gravity.

Einstein's general relativity theory of gravity expands our understanding
That gravity is a feature of the curvature of space and time,
Warping the four-dimensional geometry of space.

Space-time, like telling friends to meet you at the corner of 9th South and 3rd West
But forgetting to tell them the time to meet.
Space and time work together.

A marble rolls straight across a stretched rubber sheet.
A bowling ball placed in the middle dips the sheet downward.
Roll the marble; it circles around the heavy ball.

Massive celestial bodies like our Sun
Warp the fabric of space and time.
Space-time acts like a stretched flexible rubber sheet.
Any mass gives rise to a gravitational field.
Nothing can hide from gravity.

Earth, caught in this curved web,
Drops into orbit around the Sun.
Circling like a racecar doing never-ending laps
Around and around a NASCAR racetrack.

All mass curves space-time.
Make the Moon invisible, and
Its presence would still be felt
As our ocean tides rise and fall.

Gravity affects time as well as space.
Two clocks, one sitting highest on a staircase,
The other sitting lowest, closer to the center of the Earth,
Feeling the pull of Earth's gravity slightly more,
Ticking slightly slower—the higher clock slightly faster
In a dynamical changing space-time relationship.

Black holes are the ultimate gravity machine
Predicted by Einstein's general relativity theory,
Massively curving space and time.
Light does not escape past its event horizon.

Figuring out gravity's workings
Opened up the cosmos to space travel,
Breaking gravity's grasp on Earth.

Rockets propel into outer space,
Escaping earthly chains
At a velocity of 25,000 miles per hour.

Gravity's effects are now described and measured.
Everything responds to gravity in the same way.
Everything creates gravity in the same way.

Scientists search for subatomic graviton particles
Forging gravity's force field.
Calculated to exist, yet not detected.

Classical gravitons, singly very weak, interacting mildly.
Massless, with infinite reach, sailing at the speed of light.
Gravitons' attractive force weakens over distance.
Quantum mechanical spin of two,
Extremely difficult to detect.

New ideas modify our vision of the universe
Where other forms of detectable gravitons may exist.
Where gravity penetrates more than three dimensions (height, width, depth),
Spreading out into unseen small dimensions
That surround us everywhere.

Like hiking across a narrow log,
Walking forward or backward to safety our only one-dimensional option.
Ants crawl around the log.
With an extra dimensional option, they avoid getting stepped on.

Quantum mechanics' particles vibrate as waves.
Gravitons may vibrate, wrapping around small extra dimensions
Like string wraps a paper cylinder, creating a spool.
Limiting how gravitons vibrate
Where more than one type of graviton exists.

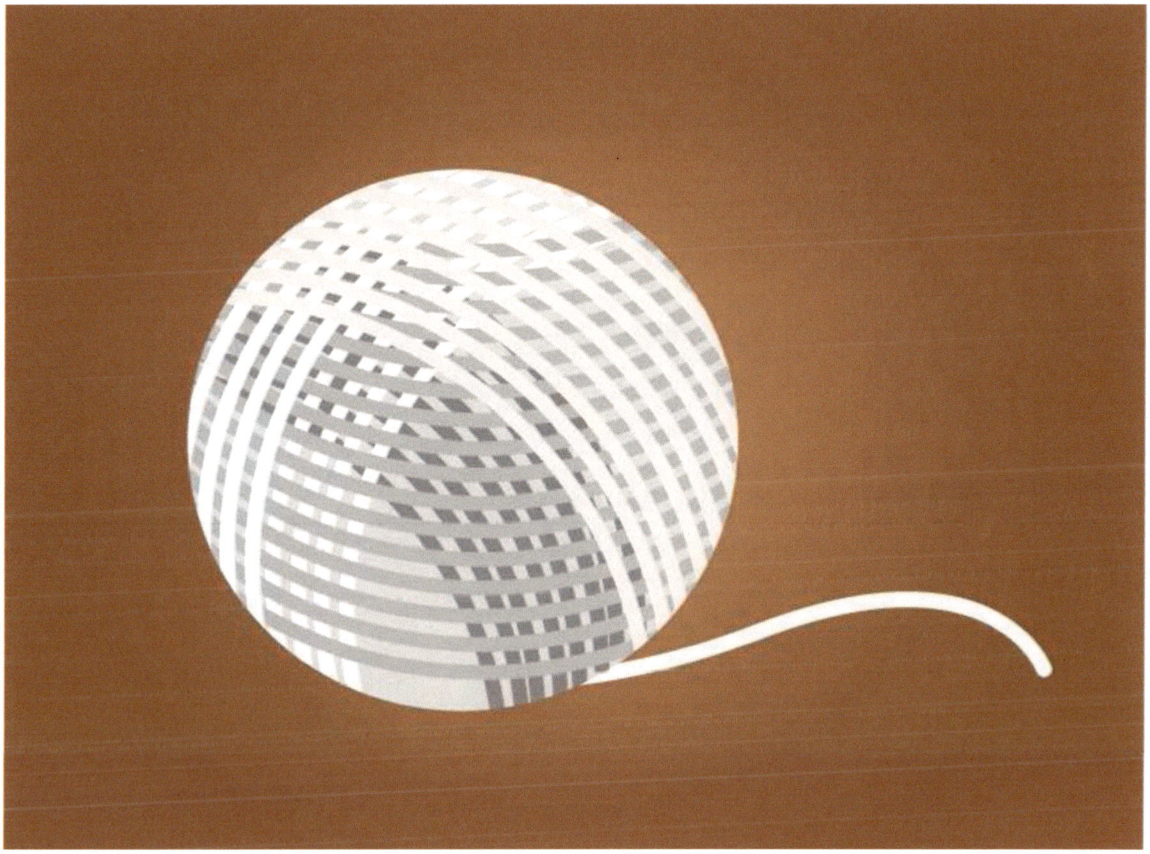

Just as various strings on the piano vibrate a very specific note.
In this world gravitons have mass on tiny scales,
On larger scales are reduced to classical massless gravitons.

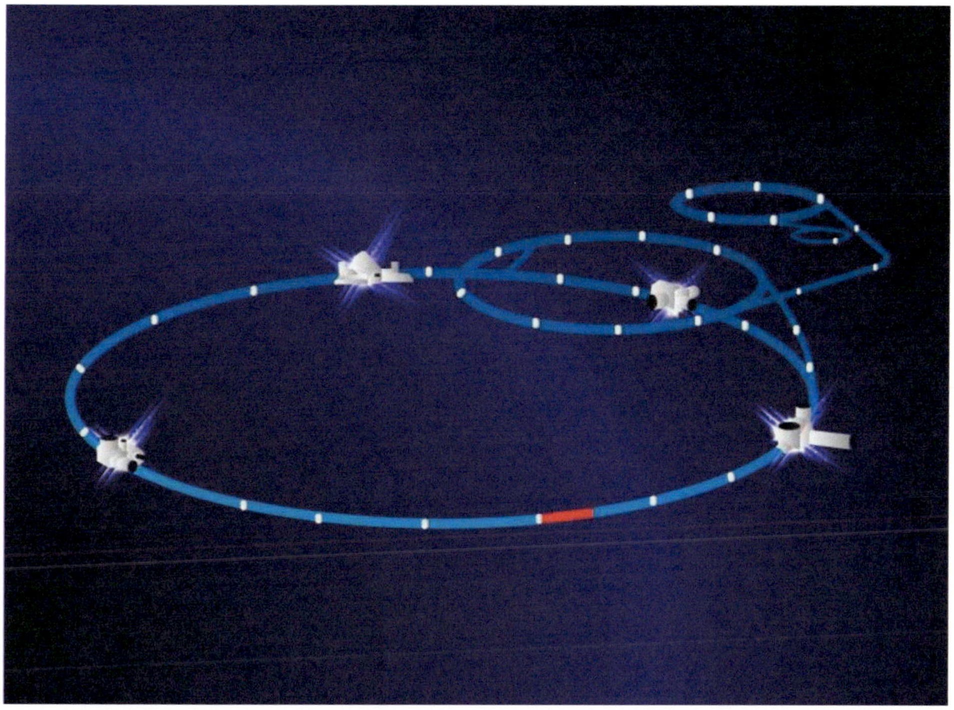

The Hadron Collider and other particle accelerators
Search for small extra dimensions,
Looking for evidence of gravitons' decay.
Gravity stubbornly resists discovery,
Like playing a game of hide and seek.

Scientific advances may one day decipher gravity's mysterious makeup,
Maybe seeing gravity itself and not just its congregating pull—
Witnessing gravity particles pulling materials together
As gravity gives form to all planets and stars
As an essential ingredient to galactic life force.

Artist conception of Gravity Probe B orbiting the Earth measuring four dimensional space-time.

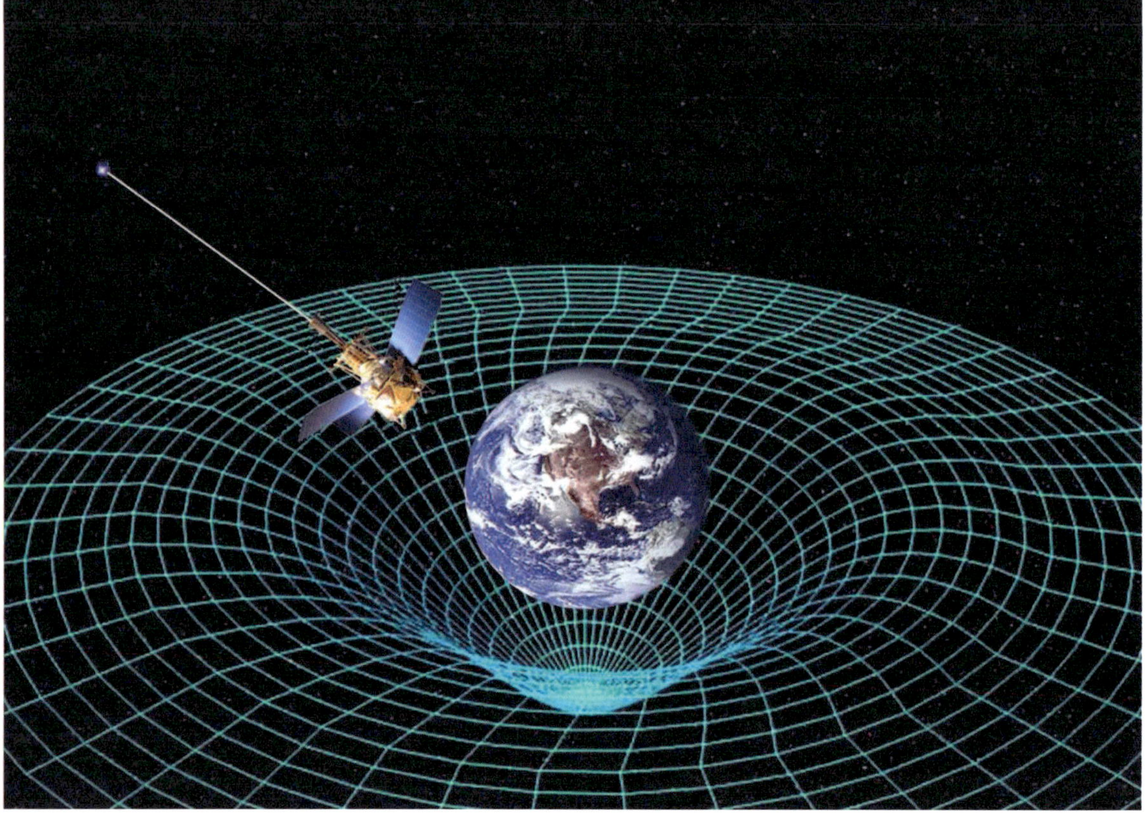

Hubble Space Telescope's image of the globular cluster Messier 56, composed of numerous stars held together tightly by gravity.

THE EXOTIC DARK MATTER STORY

Camping in Utah's Canyonlands desert,
Far away from all city lights,
Bedazzled by the brilliant nighttime display,
Black sky backdrop with splashes of glowing planets,
Shining stars, the Milky Way, and other distant galaxies.

Compared to exotic dark matter,
This cosmic beauty is quite ordinary.
Everything you see, touch, feel, smell, taste—
All galaxies, stars, planets, and people are made of ordinary matter.

All mass contains some gravity,
From stars in the sky to tadpoles in a pond.
Drifting alone in outer space,
Your gravity attracts objects.
Floating coins move slowly toward you.

Our Sun's attractive force is stronger on nearby planets.
Closer Earth orbits faster; distant Neptune slower.
Earth moves 19 miles per second; Neptune 3.37 miles per second.

Swing a bowling ball on a rope.
Rotate faster and the ball spins faster.
Rotate fast enough and you lose your grip.
The ball flies away like a cannonball.

Beyond our solar system,
Some stars within galaxies
And some galaxies within galaxy clusters
Orbit much faster than they should.

Traveling fast enough to escape and fly away,
Yet their orbital path constant for billions of years.
Avoiding rocket-like propulsion into the void.

Calculations show a shortage of visible stars, planets, gases, and dust.
Five times more ordinary matter is needed
To account for that missing mass and gravity—
Like adding two million Earths for every star in our galaxy.

Most galaxies are bound together by an invisible attractive force...
A mysterious force named dark matter.

Dark matter's halo surrounds the Milky Way.
Invisible dark matter halos surround most galaxies,
Inferred from effects on stars and gases.
X-ray images display galaxies enwrapped
Within a cloak of a hot gas cloud,
Stopped from evaporating by dark matter halos.

Gravitational lensing, massive celestial stars and galaxies bending, distorting light
Act as cosmic magnifying glasses for us to see great distances.
There's not enough visible mass present to create such lensing,
Giving more evidence of dark matter's existence.

Dark matter wields gravity
Like our Sun's gravity attracts planets,
Yet dark matter's composition is unknown.

Dark matter, not emitting, reflecting or absorbing light,
Is elusive, like an invisible person
Walking with ease through solid walls.

If you could hold dark matter in your hands,
It would have some weight—
Dissolving through your fingers.

Supercomputers simulate dark matter's evolution,
Spreading forth like a massive spider web,
Forming dense knots along the way.

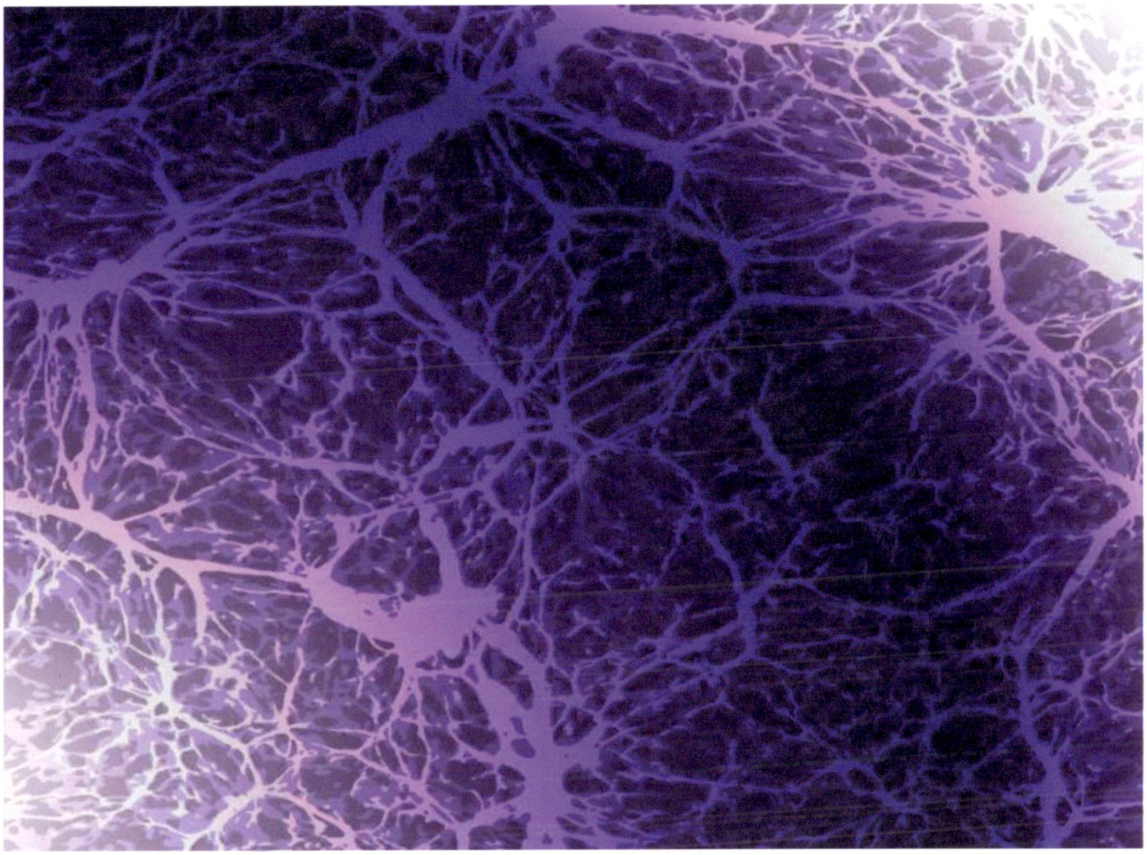

Dark matter knots match galaxies and galaxy cluster locations
As if dark matter evolved with the universe,
Possibly coming from the same place as all matter:
The universe's big bang beginning.

Dark matter clumps matter together,
Helping galaxies to form and grow.

Dark matter holds together all that is visible in the cosmos,
Acting like steel girders used to build a high rise,
Providing scaffolding beams for ordinary matter to attach to.

Dark matter may be a new kind of elementary particle,
Having mass and producing gravity,
Belonging to a family of subatomic particles,
Hardly interacting with ordinary matter at all,
And creating a quantum mechanics properties puzzle.

A wind of dark matter continuously passes through you,
Interacting just barely or not at all, undetected by your body.
Called a WIMP, or "weakly interacting massive particle."
Searches for WIMPs have brought no results yet.

String theory speaks of more dimensions around us,
Like being in a nested box of three dimensions
Inside another box and another,
Lending itself to other bubbles of a multiverse.

Super specialized and very sensitive machines
Hunt each day for dark matter.

Earth and satellite detectors look out in space for evidence of dark matter particle collisions within galaxies.

Particle colliders like the Large Hadron Collider
Try to make dark matter particles
Through ultra-high-energy particle collisions.

Deep underground, old mines now science labs
Try to detect dark matter particles
Far from cosmic particles, rays, and radioactivity interference.

Extremely sensitive crystal detectors try to detect tiny amounts of heat
Released by an incoming dark matter particle,
Producing crystal vibrations.

Similar experiments worldwide help with this search,
Using liquid Xenon and Argon, Bubble Chambers, Scintillating Crystals, and CCDs like in cameras.
Another tries to calculate seasonal variations as Earth rotates and bumps into more areas of the dark matter halo and dark matter wind.

A super-huge detector could be built to sit on Earth's surface
To try to capture these elusive dark matter particles.

DAMA (DArk Matter) scientists claim dark matter particles have been found.
A similar detector is under construction to replicate results.

Gravity wants to keep things together,
Yet expanding dark energy wants to spread things apart.
Gravity from our universe's ordinary matter alone
Could not hold the universe together.

Without adding in this dark matter,
There would be no galaxies, stars, planets, or us.
We owe our existence to dark matter.

Intrigue over this mystery sparked Dark Matter Day,
Celebrating the hunt for dark matter.

One day soon, dark matter's makeup may be uncovered,
And when found, possibly within the next decade,
It will be the discovery of the century.

The collision of galaxy clusters as shown by Chandra X-Ray Observatory and the Hubble Space Telescope show dark and ordinary matter separation.

GRAVITATIONAL LENSING

Mass and energy create gravity's power.
Planets travel a straight line
And fall slightly sideways into orbit around the Sun.
Unaware, gravity bends their journey,
Moving in a straight line in curved space-time.

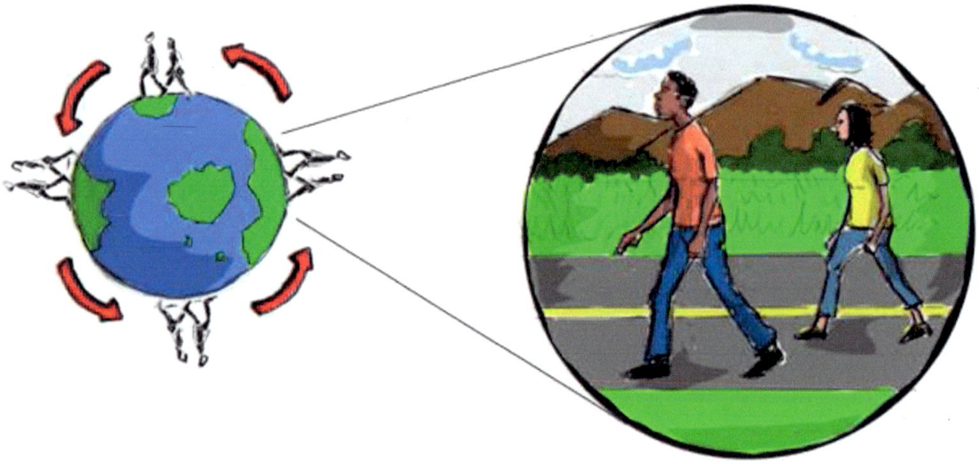

Like walking an unbending road across Earth,
Eventually ending where you started,
Walking a circular path around our curved planet.

Earth telescopes magnify heaven's light.
Pointing upward toward the night sky,
Their lenses enlarge pinpoint specks—
Captured light reveals galaxies of stars, dust, gas, and more.

Light beams travel a straight line,
Bending around massive bodies,
While gravity fields deflect light rays passing through.

Gravitational lensing can occur when massive galaxies and galaxy clusters
Bend light from a distant galaxy lying behind it,
Projecting this curved light back onto this faraway galaxy,
Giving those in its line of sight a brilliant view.

Gravitational lensing effects are like having a universe of natural telescopes,
Like spotlights lighting up a theater stage.

Dim lights of background bodies are amplified,
Distant galaxies and quasars revealed,
Outreaching existing telescopes' limited abilities.

Gravitational lensing's line of sight,
Provides magnified, shifted, split, distorted pictures—
Our views of background scenes altered.

Light converging less than perfectly,
Off-centered distorted light rays
Play tricks on our eyes
Like funhouse mirrors' warped images.

Stellar bodies hide behind galaxies;
Galaxies shield us from their view.

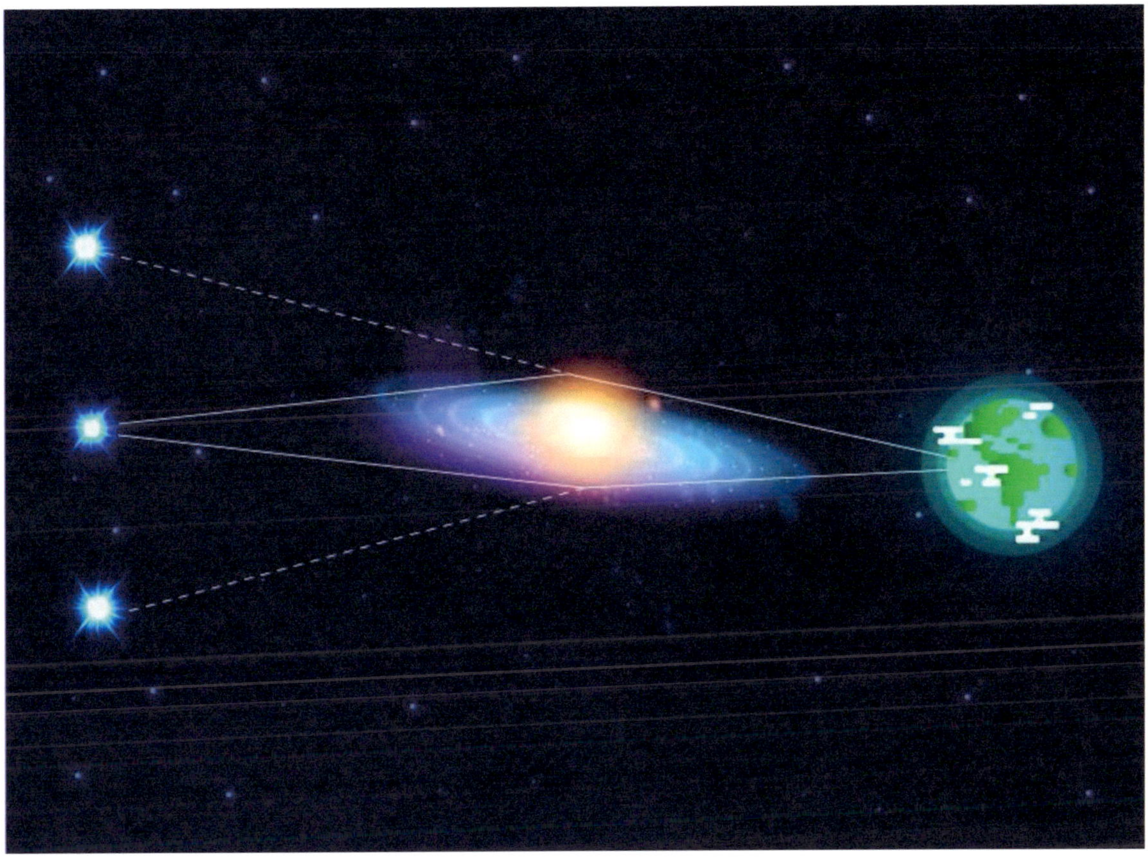

A hidden super-bright object found at a galaxy's core: a quasar.
Its light travels outward billions of years.
Light beams bend around the foreground galaxy's gravity fields.
This galaxy's huge array of stars, gas, dust, and solar systems
Act as an out-of-focus optic lens,
Projecting two quasar images, one on either side,
Reflected images from the real far distant quasar.

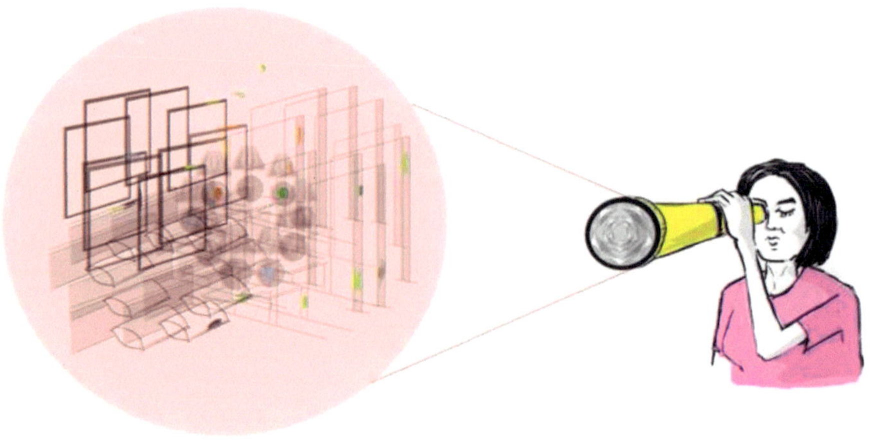

Cosmic kaleidoscopes of colors and shapes,
Like blurry images from out-of-focus camera lenses,
Their stunning artistry, are the cosmos' poetic creations.
Images such as the Einstein Cross, Huchra's lens, and Einstein ring.

A lone quasar lying squarely behind a forefront galaxy.
Gravity captures the quasar's image,
Bent light splitting into four exact pieces,
Creating an Einstein Cross.

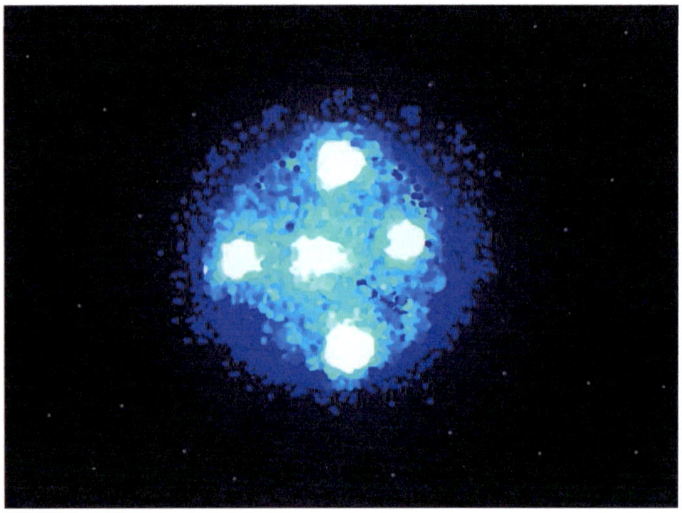

Huchra's lens at the cross's center
Of the barely visible lensing galaxy,
Billions of light-years apart,
Galaxy and quasar appear as one.

Galaxy clusters' range reaching farther than a single star,
Focusing light of a distant bright object
Into an Einstein ring around it.
Or arcs and several points
In a less than perfect alignment.

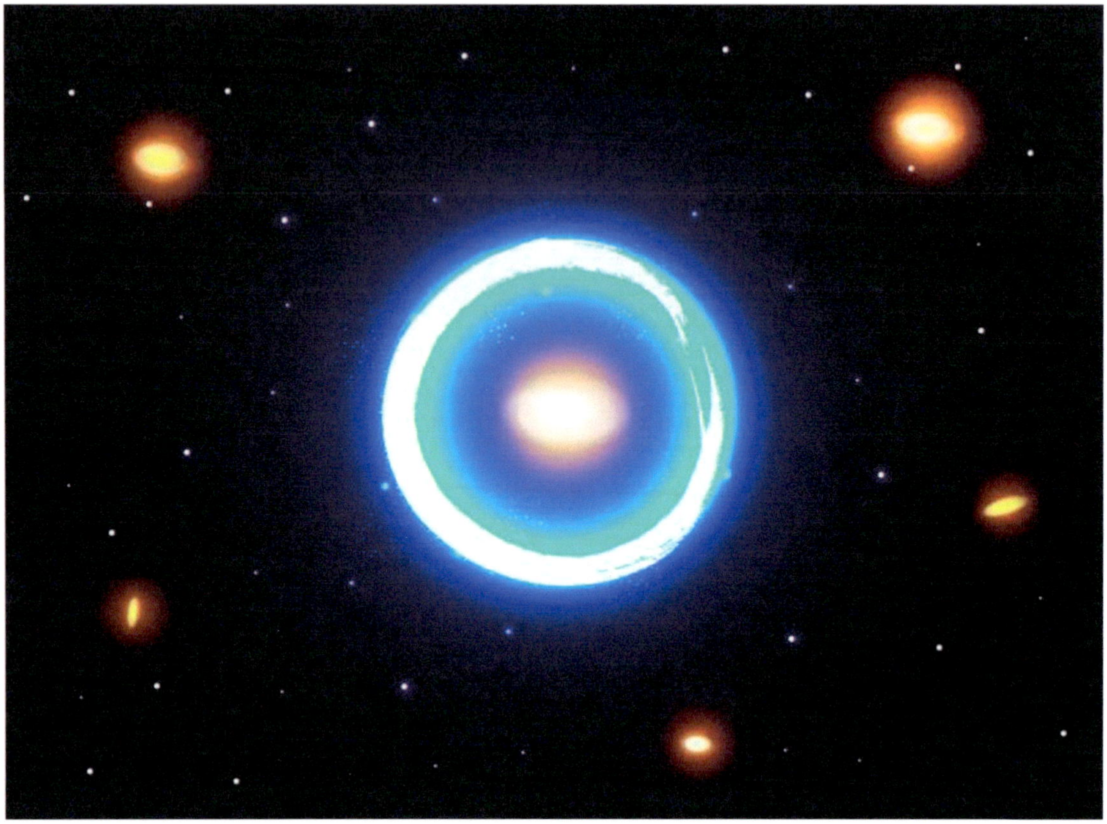

Tilt a magnifying glass at an angle
Or look through a wine glass base.
The lenses distort the view of what's behind it.

Light's path twists and warps with gravity's force,
Lensing unique images around background galaxies
Like long, thin arcs, odd blobs, and rings.

Gravitational lensing creates multiple images of a single galaxy
Or double images of the same quasar billions of light-years away.

With micro lensing's subtle brightness, variations appear over time,
Detecting dim stars and hard to find solar planets.
Planets passing by vastly affect a background star's brightness.

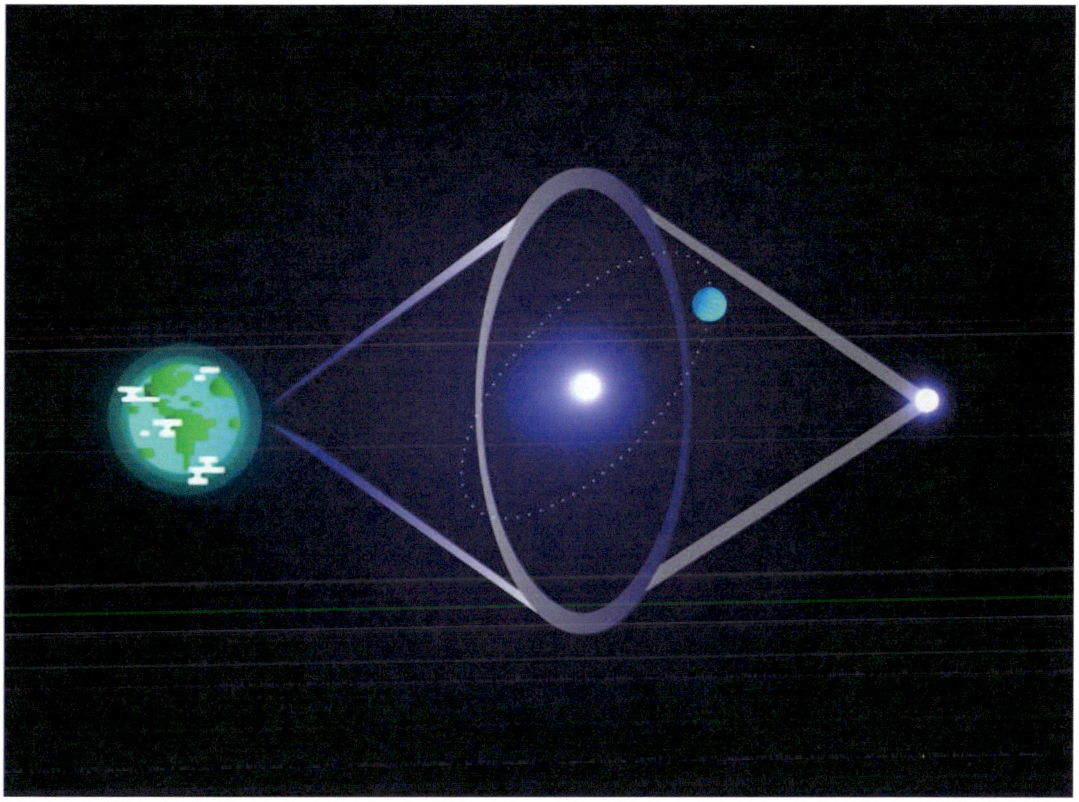

Other effects go beyond standard microlensing:
Parallax, measuring a star's movement against distant ones
As the Earth revolves around the Sun.

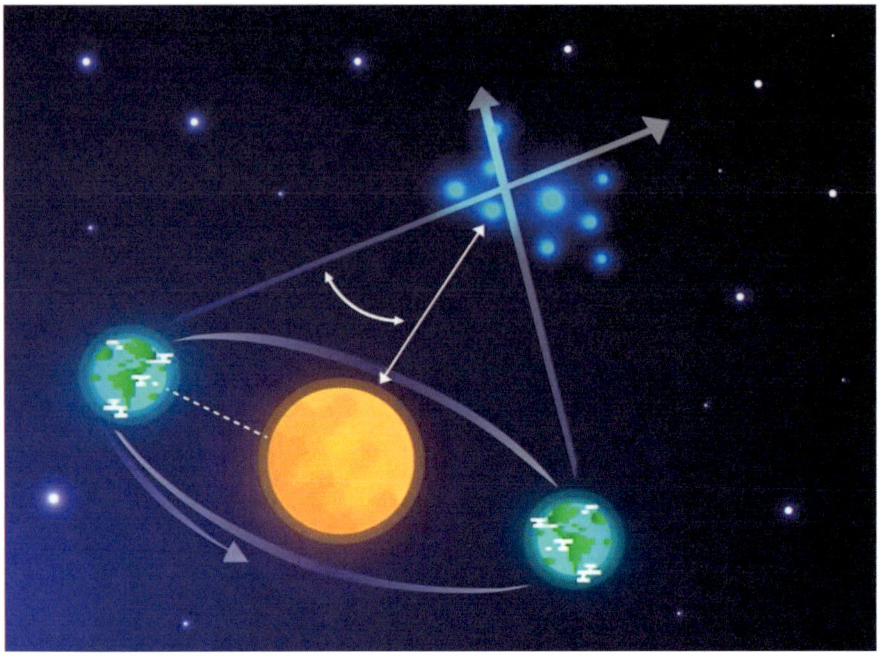

Xallarap, gravitational lensing of double star systems,
Lensing orbital motions of objects around it.

Two different effects, the reverse of each other,
Shown by reversing letters in a word design:
Xallarap is the word *parallax* spelled backward.

Looking through the curved lens of a telescope,
We see only some of the cosmos' grandeur.
Nature's gravitational lensing helps us see the un-seeable.

By peering beyond a telescope's limits,
Curved cosmic lenses provide vast views.
The universe helps us to see its majestic beauty.

An example of an Einstein Ring taken by the Hubble Space Telescope.

Young Galaxy S0901 on upper left observed by the Hubble Space Telescope through the effects of gravitational lensing.

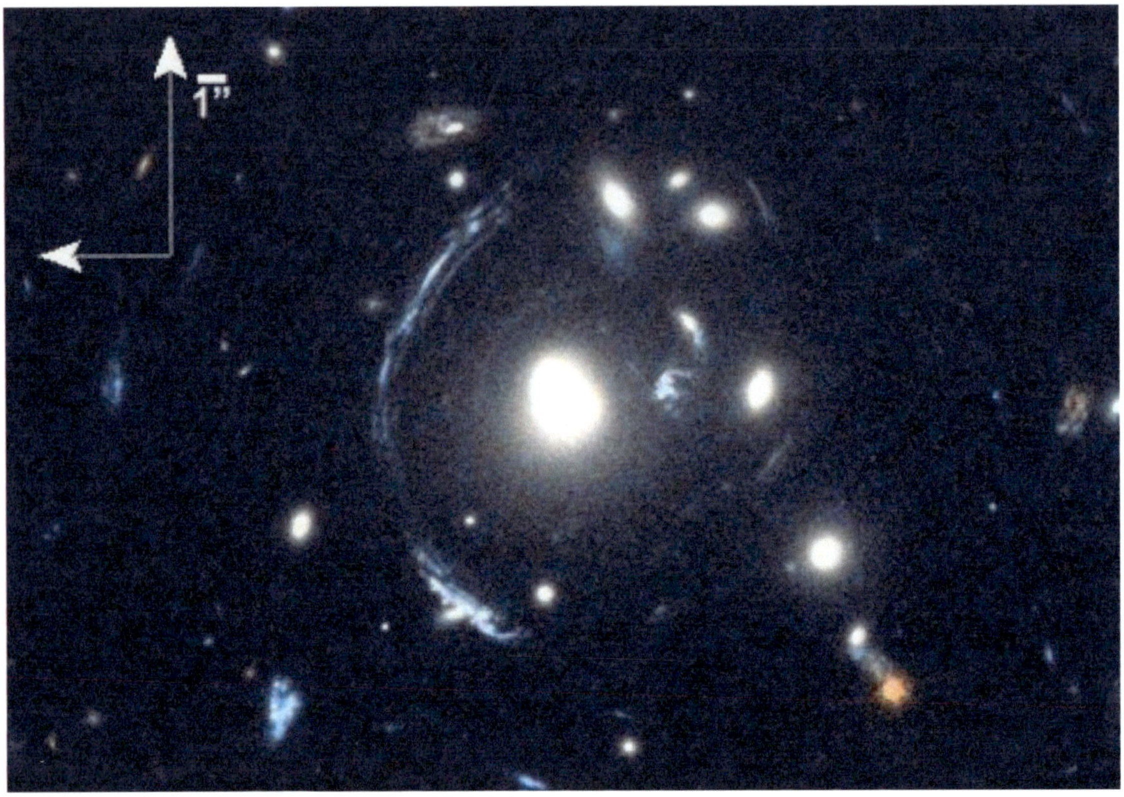

Courtesy NASA/JPL-Caltech.

GRAVITATIONAL WAVES FIRST DISCOVERY

Once upon a time, 1.3 billion years ago,
In a very distant galaxy,
Two black holes orbiting around each other
Drew dangerously closer and closer together.

Swirling one around the other,
Their disturbance like a powerful ocean storm,
Generating great ripples in space,
They finally collided in a brief mega burst—
Cataclysmic cosmic crash, two black holes merged into one.

Two black holes, solar masses of 36 and 29,
Smashing into one, totaling 62 solar masses.
In two-tenths of a second releasing three Suns' worth of energy.

Fifty times more power briefly generated
Than all stars in the known universe
Was released as gravitational waves, ripples in the fabric of space-time itself.
Like wakes speedboats create when they move through water.

Energy exploded across the universe—
Ripples sent outward in every direction, traveling the speed of light,
Their orbital strength lessening over distance trekked
Like pond ripples from a stone breaking its surface.

Gravitational waves stretch space and everything out in one direction,
Squeezing space and everything back in the opposite direction,
Reversing motion side to side, up and down,
Like a rhythmic dance flowing out and back in continual motion.

Gravitational waves passing through Earth and us,
Weakened greatly over their billion-year journey,
Expanded and contracted everything very, very slightly,
Stretching and compressing us by a decimal point, twenty zeros, and then a one. (.000000000000000000001)

Swing your arms around each other.
You create gravitational waves,
Though much too weak to detect.

Earth orbiting the Sun makes gravitational waves,
Their energy too weak to be felt.
Strong waves threaten our solar system's stability.
Like gentle waves rolling softly upon the beach
Versus a tsunami's destructive force.

Einstein's math formulas predict gravitational waves,
One hundred years pass for direct detection.
Astronomical drama required, like two black holes colliding,
Generating waves colossal enough to detect on Earth.

September 14, 2015: a historic day!
The first physical gravitational-wave discovery,
Taking four decades for the project's technology to advance
To finally capture one.

Two advanced Laser Interferometer Gravitational-Wave Observatories
(LIGO), Louisiana and Washington State,
Captured gravitational waves' inconceivably small movements.

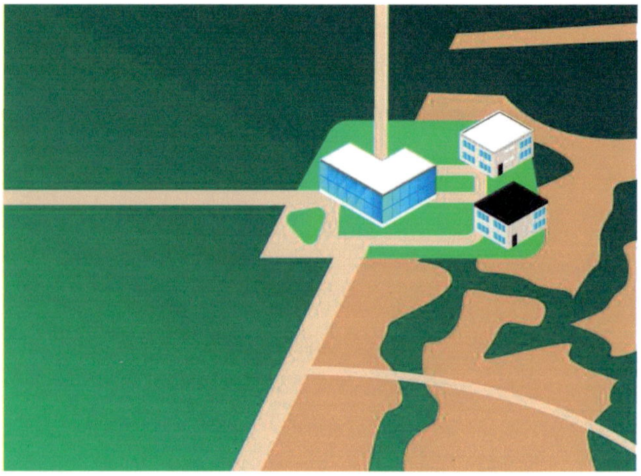

LIGO's L-shaped perpendicular
arms span 2.5 miles each.
Rounded concrete covers protect
vacuum tubes within.
Laser beam light, split at the beam
splitter, shoots down each arm,
Bouncing off extremely sensitive
mirrors.
Photodetectors compare laser
beams' changing lengths.

Mirrors suspended by ultra-strong glass strands
Are hung by four multiple pendulum layers.
Special materials and mechanisms minimize seismic jiggles,
And gravitational waves move mirrors while detectors measure differences.

Mirrors are unmoved by passing vehicles, wind, or shaking earth.
Like a yoyo with the string let out, move your hand, the yoyo stands still.

Gravitational waves compressing space-time ever so slightly along one arm
Expand space-time along the other arm.
Reversing, expanding, compressing back and forth as gravity waves pass by.
The measured length difference about one one-thousandth of the width of a proton inside the nucleus of an atom.

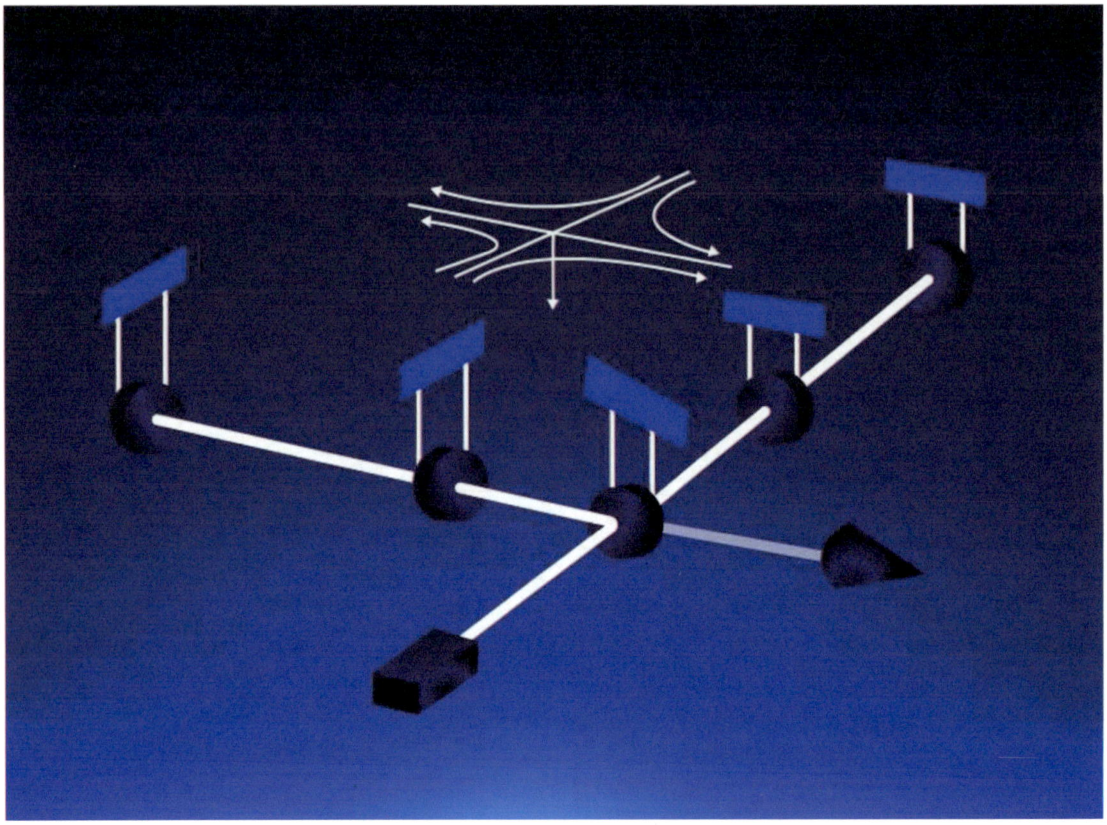

Take one meter, 3 feet.
Divide by 10,000, a human hair width,
Divide by 100, the wavelength of light,
Divide by 10,000, an atomic diameter,
Divide by 100,000, a nuclear diameter,
Divide by 1,000,000, LIGO's sensitivity.

About 5 million million atoms fit on the head of a pin.
If you enlarge the atom to the size of a football field,
Its proton would be the size of an ant.
Split the ant into 1000 tiny pieces, and
One piece represents the width difference caused by the gravitational-wave.

Gravitational waves converted into sound,
Listening to changes in amplitude and frequency.
The sound of two black holes joining
As a chirp sounds two-tenths of a second long,
Starting two octaves below middle C, then ending at middle C.

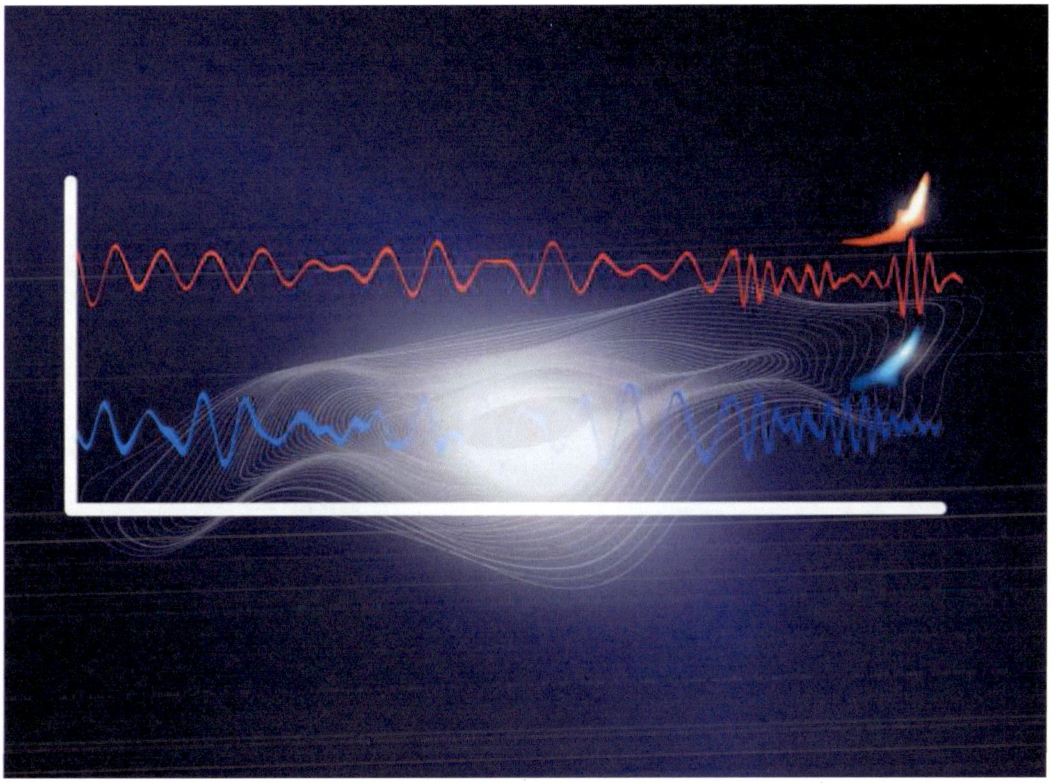

Since LIGO's historic first, more detectors work in sync.
VIRGO/Italy, KAGRA/Japan, GEO/Germany, soon LIGO/India.
LIGO's gear design inventively fine-tuned,
Squeezing light devices, stronger laser beams, longer arms, lighter materials,
Evolving with advanced technology.

LIGO observes more black hole collisions,
Neutron star collisions, and supernovae collapses.

Gravitational waves emit from supernovae's very center,
Released from collapsed cores, supernovae explosions,
LIGO voyages past where light cannot travel.

One hundred and thirty million light-years ago,
A pair of neutron stars, compact, dense remnants of dead stars,
Drew closer and closer together.
Sending out gravitational-wave energy,
Colliding, bursting with gamma rays.

Detected by LIGO and VIRGO, their first detection of colliding neutron stars
Took place on August 17, 2017.
Waves also converted to sound, a much longer chirp.

The Fermi Gamma-Ray Space telescope viewed these gamma rays
As light emitted over the entire electromagnetic spectrum.
Light, radio wave, x-ray telescopes caught the brilliant show of crashing
neutron stars.

Involving seventy observatories, Earth and space detectors,
Multi-messenger astronomy, worldwide collaboration for universal goals.

Neutron star collisions, long thought to produce gamma-rays,
Remained unproven until LIGO.
LIGO also proved neutron star collisions create gold and platinum.
The gold in your ring, likely billions of years old.
Neutron stars violent explosions creature nature's own refining furnace.

Gravitational-wave pioneers Rai Weiss, Barry Barish, and Kip Thorne
Were awarded the 2017 Noble Prize in Physics
For creative ingenuity and dedication.

LIGO's gravitational-wave detection made possible
A brand new field of astronomy.

Space gravitational-wave detector, LISA, will launch in the 2030s.
Its mission: detecting gravitational waves farther back in time,
Passing through the cosmic microwave background's opaqueness
To the very beginning of our universe.

Exploring big bang beginnings,
Gravitational waves traveling through untouched,
Penetrating hot, dense, earliest moments.

LIGO offers a new way to observe the universe.
LIGO/LISA will let us see all the way back to the very beginning.
Our universe's unexpected wondrous wonders and super marvels are soon coming our way.

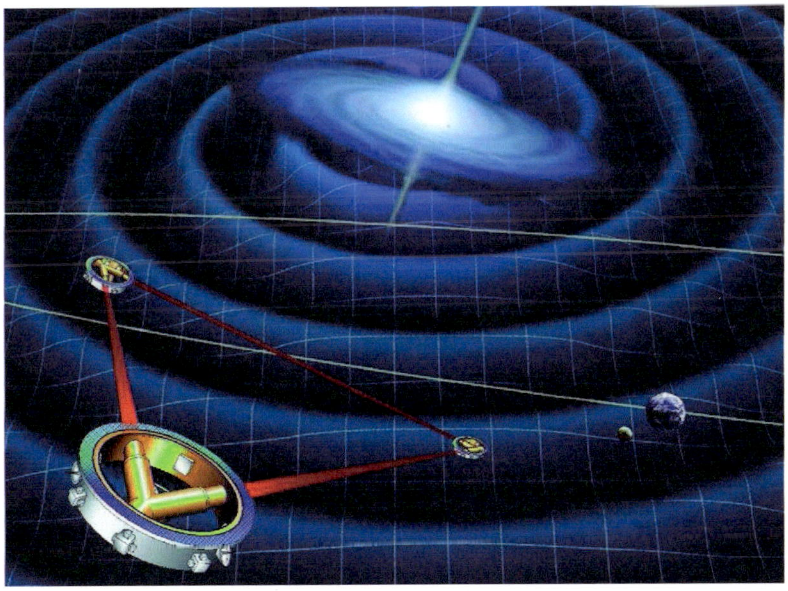

Artist's rendering of the future LISA Space Mission,
consisting of three spacecrafts.

A supercomputer simulation of merging black holes and the resulting gravitational waves.

COLLIDING NEUTRON STARS

Twinkle, twinkle, neutron star,
Oh what wonder that you are,
Up above the world so dense,
Remnant of a massive star.

Born from gravity's collapse,
Twinkle, twinkle, neutron star.

Then two neutron stars collide
In a galaxy up high,
Tearing each other to mere shreds,
Captured both in sight and sound.

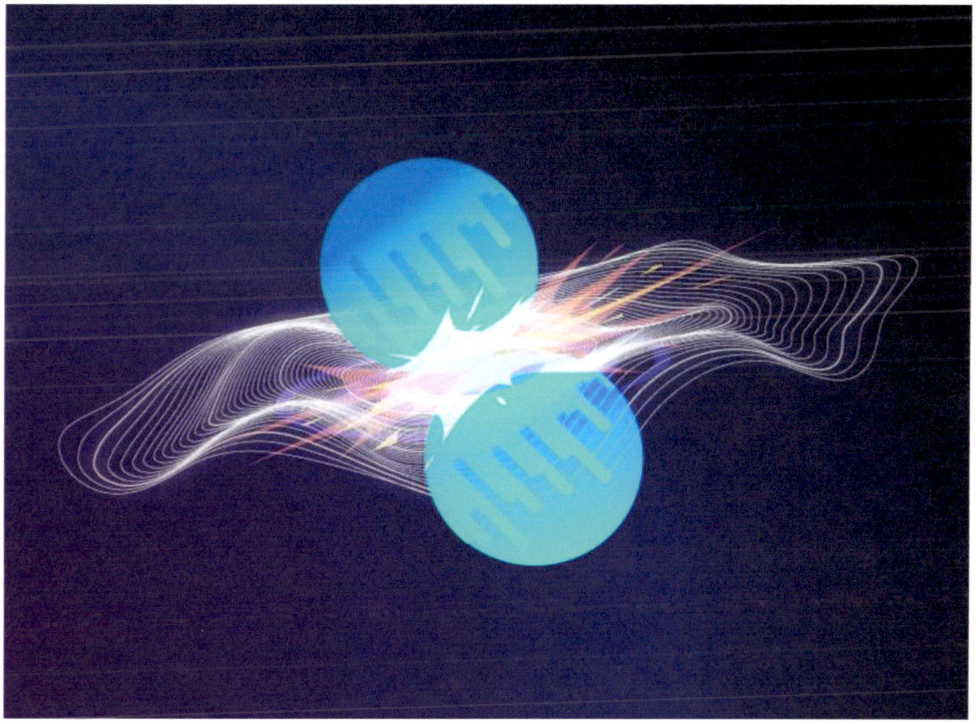

This now being a first,
Heard by LIGO and the rest.

Sounding for 100 ticks,
Much more than a black hole's short chirp—
Waves that move both space and time.
Gravity makes rippled lines.

Sight and sound together flow;
Heaven's own fireworks show.

Seen by telescopes on high,
The violent crash spews gamma rays.
Such a light show in the sky,
Matching with surrounding sound.

Making heavy elements:
Gold and silver and much more.

What remains there is not known.
It could now be a new black hole
Or a short-lived neutron star.
Maybe one that's living long.

Crashing, crashing neutron stars,
What have you two now become?

This illustration shows a cloud of debris from two neutron stars just before they collided in galaxy NC 4993, 130 million light years from years from Earth.

BLACK HOLE BASICS

Massive black holes lurk
Within the centers of many galaxies,
Capturing, devouring everything within their reach.

Supermassive black hole, Sagittarius A*,
Resides at the Milky Way's center,
Totaling 4 million solar masses.

One solar mass equals about 333,000 Earths—
A hollow Sun with Earths filled to the top.
Like taking an empty jar, stuffing it with marbles.

Mass: how much matter something is made of.
Planets, stars, you, your friends, your backpack, all have mass.
You're made of bones, fat, blood, water, tissues, muscles, organs,
Remaining the same wherever you are:
Earth, Moon, the Space Station, or Mars.

Only your weight changes throughout the universe,
Weighing more on Earth than on the Moon.
Gravity's pull is greater on Earth, less on the Moon.
What weighs 120 pounds on Earth
Becomes 20 pounds on the Moon.

Black holes are highly compact, extremely dense.
Black holes are the mass of our Sun
But span only 2 miles across—
Like compressing our Sun into a large college campus.

Imagine junkyard cars flattened by a crusher machine.
All metal, plastic, rubber, and glass remains.
The cars are crushed, now thinner in form.

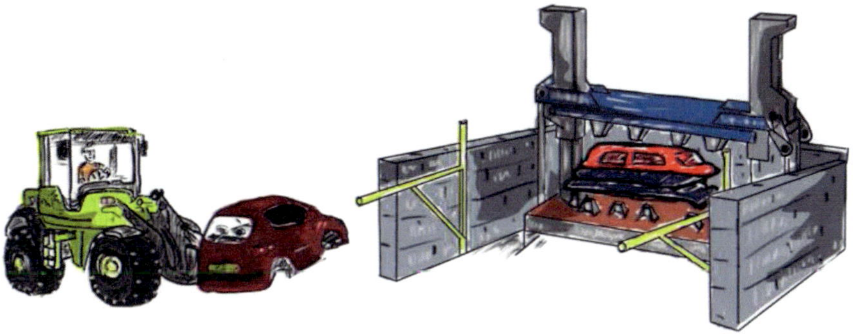

Black holes are called black because all light is pulled in at their borders,
Exerting dominion over all in their path.
"Not even light" escapes their grasp.

Called *holes*, their gravity sinks through space-time fabric,
Creating bottomless holes in space.

Keck infrared telescopes watch the stars
In the middle of the Milky Way,
Orbiting around a dark center.
We measure our black hole's mass
From these stars' orbits.

Black holes, small and huge, are found throughout the universe.
In-between sizes are yet to be discovered,
Yet math formulas predict their existence.

Smaller stellar black holes
Are formed as stars explode.
As their lives ends,
Black hole remnants are left behind.

Supermassive black holes abound.
Exactly how they formed remains an unanswered question.

Black holes contain boundaries.
An event horizon with no surface.
This is unique to black holes, as everything we know has a surface.
Like scuba diving deep depths under dark seas,
Water engulfs you and the surface disappears.

Event Horizon's array of telescopes,
Its high resolution Very Long Baseline Interferometry (VLBI),
Locates the inner orbit of supermassive black holes.

Telescopes begin to measure black holes' spins,
Yet a black hole's electromagnetic charge can't be measured by current devices.
Mass, spin, and charge measurements define a black hole.

April 11, 2019, history was made by Event Horizon telescopes
When a first-ever black hole image was captured
At the center of Galaxy M87.
Outlining its silhouette,
Glowing gas falling into the event horizon.

If you crossed over the event horizon
You would appear to hover,
Becoming fainter and fainter
Until as a faint red glow
You disappeared from sight forever.
From the outside, we are unable to peer within.

If you entered feet-first into a smaller stellar black hole
Gravity would stretch you out,
Pulling on your feet more than your head,
Squeezing your sides together.

Squashing you in sideways
While pulling you apart lengthwise,
Forcing you to spaghettify
Into angel hair pasta size strands.

Row a wooden boat near a massive waterfall.
Above the falls if you paddle fast enough, you'll get away.

Yet cross too near the edge,
The waterfall's power will grab you,
Plunging you down its vertical drop,
Your boat splintering apart.

In our four-dimensional universe,
Space-time is curved by the matter in it.
A heavy object dips a rubber sheet,
Smaller objects follow a curved path around it.

Place heavier and denser
Weights on the sheet,
And the sheet depresses more and more.

At a critical weight and size,
A bottomless hole in the sheet is made.
Objects and particles fall into the hole,
But nothing gets out.

You could possibly survive travel through a hulk-size supermassive black hole,
Crossing the event horizon with ease,
Traveling through a wormhole passage from one universe to another,
Or back to another place in the same universe.

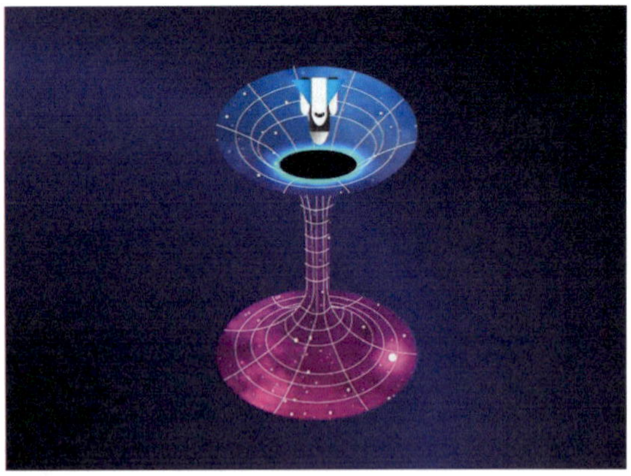

Supermassive black holes are everywhere.
Most today are dormant,
Snacking now and then.

Yet when they feed,
They form an accretion disk of gas,
A quasi-stellar object, a Quasar,
That looks like a star, emitting light.
Black holes, when feeding, shine.

The era of quasars is over,
Peaking several billion years ago.
Black holes can be active,
And when they feed a lot, they are very bright.
Some very powerful ones are still out there.

The Large Hadron Collider particle accelerator,
The world's largest machine, is a seventeen-mile ring
Smashing particles together, traveling in opposite directions,
Hoping to create a mini black hole.

Black holes' wonderment abounds
As we explore their growth in tandem with galaxies,
Uncovering hidden connections.

With galaxy collisions, black holes should collide also,
Supermassive black holes becoming as one.
A pair merging recently detected became another historic first.

Improved technology and telescopes,
With much effort and ingenuity,
Will uncover more cosmic answers.
Our universe is not obligated to reveal answers to our questions.

Event Horizon Telescope Collaboration's first image of a black hole at the center of galaxy M87.

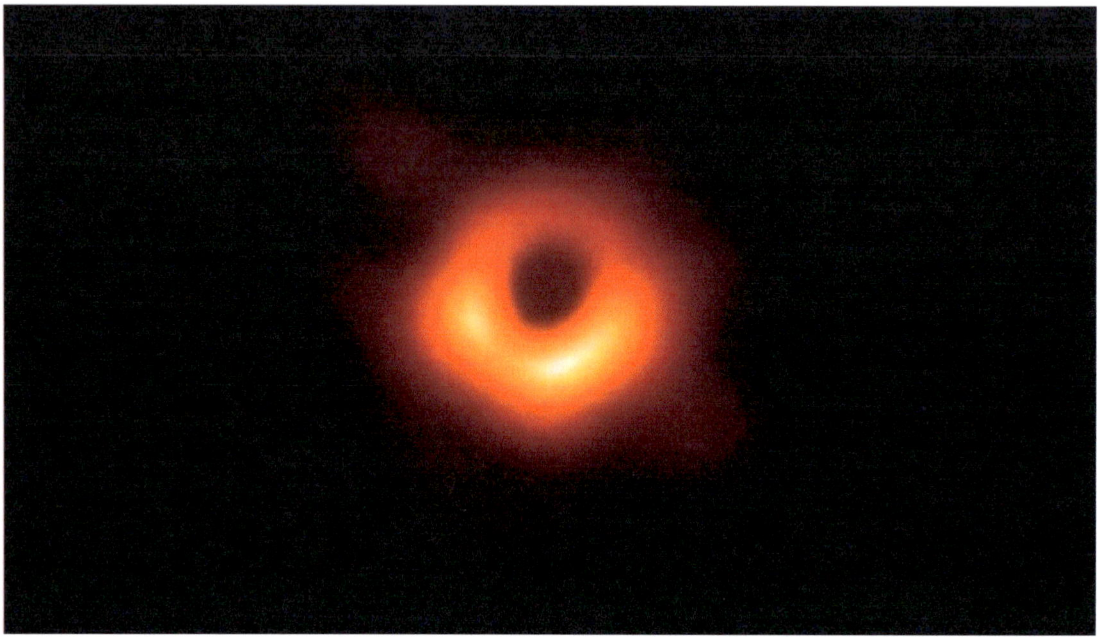

Courtesy NASA/JPL-Caltech.

REFERENCES

1: The Big Bang Theory of Cosmic Evolution
Advexon Science Network (March 3, 2019). *What is Space-Time and How it Works* documentary.
deGrasse Tyson, Neil (5/2/2017). *Astrophysics for People in a Hurry*. W.W. Norton and Company.
Green, John (9/17/2014). *The Big Bang: Crash Course. Big History #1, Crash Course.*
Howell, Elizabeth (8/29/2016). *Cosmic Microwave Background: Remnant of the Big Bang.* Space.com.
Howell, Elizabeth (11/7/2017). *What is the Big Bang Theory?* Space.com.
Impey, Chris, (2/22/2011). *Origin of the Universe. University of Arizona Lecture Series.*
Keus, Venus, (2/19/19). *Three Ways the Universe Could End.* Ted-Ed.

2: Dark Energy
Dark Energy Survey (2019). DarkEnergySurvey.org
deGrasse Tyson, Neil (May 2, 2017). Astrophysics for People in a Hurry. W.W. Norton and Company, New York/London.
Filippenko, Alexei (Oct 16, 2015). Exploding Stars and the Accelerating Universe. American Academy of Arts and Sciences.
Kirshner, Robert (2018). Exploding Stars, Dark Energy and the Accelerating Universe, Buhl Lecture.
Kirshner, Robert (September 11, 2015). *What is Dark Energy? Earth Sky Voices* in Science Wire, Space.
Taylor Redd, Nola (May 1, 2013). What is Dark Energy? Space.com.
Taylor Redd, Nola (May 12, 2012). Edwin Powell Hubble: Biography. Space.com.

3: Gravity
Advexon Science TV (Feb. 10, 2016). The Force of Gravity/What is Gravity Documentary.
Al-khalili, Jim; Smith Vicki; Appleby, Graham (2017). The Amazing World of Gravity (Video).
Carroll, Sean (2007). Dark Matter, Dark Energy: The Dark Side of the Universe. The Great Courses Plus Online
deGrasse Tyson, Neil (July 27, 2017) Comic Queries: What is Gravity? Podcast Affiliates Network.
Greene, Brian (2017). The Fabric of the Cosmos, Season 1 Ep 1, What is Space? Nova.
Greene, Brian (2017). Quantum Reality: Space, Time, and Entanglement. World Science Festival.
Koberlain, Brian (Feb. 6, 2015). What is Gravitational Lensing? (W/Video). Universe Today, phys.org.
Lincoln, Don (2014). What Are Gravitons? The Nature of Reality. Nova Online

Rosen, Raphael (May 5, 2014). How Cosmic Lenses Reveal the Depths of Deep Space (Video). Space.com
Schumacher, Benjamin (2013). Black Holes, Tides, and Curved Space-time: Understanding Gravity. The Great Courses Plus Online

4: The Exotic Dark Matter Story
Bauer, Dan (October 7, 2018). What is Dark Matter and Why Does it Matter? Fermilab.
deGrasse Tyson, Neil (May 2, 2017). Astrophysics for People in a Hurry. W.W. Norton and Company.
deGrasse Tyson, Neil (Feb. 14, 2017). Documentary- Dark Matter- Change Face. Nova Science NOW.
Taylor Redd, Nola (May 1, 2013). What Is Dark Matter? Space.com

5: Gravitational Lensing
Berman, Bob (Sept. 18, 2015). Weird Object: The Einstein Cross. Astronomy.com magazine
deGrasse Tyson, Neil (May 2, 2017). Astrophysics for People in a Hurry. W.W. Norton and Company.
Greene, Brian (January 4, 2017) NOVA: The Fabric of the Cosmos: What is Space? Season 1, Episode 1. NOVA.
Hofmeister, Caitlin (June 3, 2014). What is Gravitational Lensing? SciShow Space.
Krauss, Lawrence, (January 24, 2014). Gravitational Lensing. Science Net.
Lincoln, Don (June 24, 2015). Gravitational Lensing, Fermilab.
O'Dowd (June 15, 2016). The Strange Universe of Gravitational Lensing. Space PBS Space-Time, Digital Studies.
Sutter, Paul (March 16, 2018). Nature's Lens: How Gravity Can Bend Light Like a Telescope. Space.com

6: Gravitational Waves First Discovery
References:
Adams, Alan (3/10/16) What the Discovery of Gravitational Waves Means. Ted Talk.
Bergmann, Jonathan (Apr. 16, 2012) Just How Small is an Atom, Ted-Ed.
Blair, David (Feb. 11, 2016). Gravitational Waves Discovered: The Universe has Spoken. The Conversation.com.
Crockett, Christopher (Feb. 11, 2016). Black Holes Explained. Science News. Sciencenews.org
deGrasse Tyson, Neil (5/2/2017). Astrophysics for People in a Hurry. W.W. Norton and Company.
Dept. of Physics (10/22/2017). Q&A Atoms on a Pin Head. Univ. of Illinois Urbana-Champaign.
Francis, Matthew R. (July 2015). The Search for Gravitational Waves. Air and Space Magazine. Airspacemag.com.
Greene, Brain (June 22, 2016). Gravitational Waves: A New Era of Astronomy Begins.

World Science Festival.
Guthman, Les (December 10, 2019). LIGO Documentary, Episodes 1-9. Ligo.caltech.edu
LIGO Caltech (July 10, 2019) What is LIGO? LIGO.Caltech.edu Website.
Wall, Mike (Feb. 11, 2016). Epic Gravitational Wave Detection: How Scientists Did It. Space.com

7: Neutron Stars Collide
Norton, Perry Anne (1/7/2018) Star Crash: The Explosion that Transformed Astronomy. Cosmic Journey Season 1 Episode 3.

8: Mind-Boggling Black Holes
Filippenko, Alex (4/28/2011). Dark Energy and the Runaway Universe. Distinctive Voices, The Beckman Center, Insight on Science, Technology, and Medicine. A program of the National Academy of Sciences, National Academy of Engineering and Institute of Medicine.
Ghez, Andrea (March 3, 2016). The Monster at the Heart of Our Galaxy. The Royal Society.
Greene, Brian (June 2, 2018). Darkness Visible: Shedding New Light on Black Holes World Science Festival.
Hawking, Stephen Hawking (2008). Lecture: Out of a Black Hole. USC College of Letter, Arts, and Sciences.
Kollmeier, Juna (April 27, 2015). At the Edge of Reason: The Black Holes in the Universe. Carnegie Observatories Astronomy Lecture Series.
Lutz, Ota (April 19, 2019), How Scientists Captured the First Image of a Black Hole. NASA JPL Education
Taylor Redd, Nola (April 9, 2015). Black Holes: Facts, Theory and Definition, Science.com

ABOUT THE AUTHOR

Deborah S. Wallace is a Licensed Clinical Social Worker. She is passionate about helping others succeed from a strengths-based perspective. In addition to social work, Deborah has many hobbies such as rock hounding, mountain biking, and creative writing.

Her desire to inspire youth to explore the sciences and expand their universal perspective on a macro and micro level is the inspiration for this book, *A Universe of Stories.*

Her fifth-grade teacher, Mrs. Audrey Greenberg, inspired her when she was struggling to like science. Mrs. Greenberg encouraged Deborah to participate in the school's science fair that year and assigned her the topic of light refraction, reflection, and diffraction. Deborah wondered why her science teacher would give her a topic that seemed so boring and irrelevant to her life.

Yet through her research on the topic, a whole new world opened up for her. Much to her surprise, Deborah won second place in the science fair that year. She gained a fascination for science topics as well as a love for lifelong learning. Deborah hopes this book will help inspire other kids, and that all kids will have a Mrs. Greenberg in their lives that encourages them and believes in their potential.

Made in the USA
Monee, IL
05 November 2026

67945042R00062